Virtual International Mammalian Synthetic Biology Workshop (mSBW 2021)

Online
19-21 July 2021

ISBN: 978-1-7138-4458-7

Printed from e-media with permission by:

Curran Associates, Inc.
57 Morehouse Lane
Red Hook, NY 12571

Some format issues inherent in the e-media version may also appear in this print version.

Printed with permission by Curran Associates, Inc. (2022)

For permission requests, please contact AIChE
at the address below.

AIChE
120 Wall Street, FL 23
New York, NY 10005-4020

Phone: (800) 242-4363
Fax: (203) 775-5177

www.aiche.org

Additional copies of this publication are available from:

Curran Associates, Inc.
57 Morehouse Lane
Red Hook, NY 12571 USA
Phone: 845-758-0400
Fax: 845-758-2633
Email: curran@proceedings.com
Web: www.proceedings.com

TABLE OF CONTENTS

1-SENSORS & MULTI-INPUT LOGIC

2-FOUNDATIONAL ADVANCES

3-APPLICATION IN STEM CELLS

4-KEYNOTE WENDELL LIM

5-IMMUNOTHERAPY AND CELL THERAPY

6-GENE REGULATION

POSTERS

Feedback responsive cell factories for biomanufacturing

- Conference: Mammalian Synthetic Biology Workshop
- Year: 2021
- Proceeding: 2021 Virtual International Mammalian Synthetic Biology Workshop (mSBW)
- Group: General Submissions
- Session:

  Sensors and Multi-input Logic

- Time:

  Monday, July 19, 2021 - 10:45am-11:15am

Authors:
Segatori, L. - Presenter, Auburn University

Mammalian cells rely on complex and highly dynamic gene networks to maintain cellular homeostasis in response to environmental stimuli and intracellular signals. Efficient cellular reprogramming thus requires integration of orthogonal components for cell engineering with innate cellular networks through feedback control systems. We explored the use of post-translational tools for superior feedback regulation of dynamic behaviors. Specifically, we demonstrated efficient detection and manipulation of the signaling network mediating the cellular stress response for the design of high producing cell lines for protein manufacturing and for the development of cell therapies for sustained protein secretion.

Our approach is based on the use of the NanoDeg (ACS Synth Biol. 2018 Feb 16;7(2):540-552) – a bifunctional system that mediates proteasomal degradation of a target protein with high specificity and exquisite control over rate of decay. Iterations of experimental measurements and predictive modeling resulted in the design of a platform technology for quantitative, multiplexed profiling of gene expression signatures of the stress response with high sensitivity and dynamic resolution of the stimulus causing proteotoxic stress (*Nature Chem Biol.* 2020 May;16(5):520-528).

Current work is focused on deploying this technology for manipulating stress signaling pathways in response to feedback signals generated at different stages of protein production, with the overall goal to engineer cells for enhanced and sustained expression of recombinant proteins. This talk with describe strategies for engineering cells based on feedback signals that enhance the stress attenuation response induced in response to proteotoxic stress and delay ER-stress induced apoptosis with respect to activation of the stress attenuation response.

Polymeric encapsulation of bacterial biosensors enables co-culture with mammalian cells

- Conference: Mammalian Synthetic Biology Workshop
- Year: 2021
- Proceeding: 2021 Virtual International Mammalian Synthetic Biology Workshop (mSBW)
- Group: General Submissions
- Session:

  Sensors and Multi-input Logic

- Time:

  Monday, July 19, 2021 - 11:15am-11:45am

Authors:
Polizzi, K. M. - Presenter, University of Exeter

Synthetic biology has enabled us to harness the ability of living systems to sense their environment and respond to changes to develop genetically-encoded biosensors for a variety of applications including biomanufacturing, healthcare, and as orthogonal control of cells. From a biomanufacturing standpoint, biosensors offer increased sensitivity and specificity for detection of key metabolites compared to traditional detection methods. However, one of the remaining challenges is whether the sensing elements should be placed inside the producing cells (which diverts cellular resources from bioproduction towards sensing) or whether dedicated biosensor cell populations could be created and cultured alongside the producer cells. In the latter case, the growth and division of the biosensor cells must be limited or they will overgrow the producer cells, leading to an overall loss of productivity from the culture. To tackle this challenge, we have been exploring strategies involving encapsulation of biosensor cells in materials that allow free diffusion of metabolites to the biosensor cells, but limit their growth within the culture.

In this talk, I will introduce LAMPS (Living Analytics in a Multilayer Polymer Shell) and demonstrate their application to the co-culture of a bacterial whole-cell biosensor for L-lactate with suspension-adapted mammalian cell lines producing a monoclonal antibody. LAMPS are composed of an inner layer of alginate containing the biosensor population, surrounded by one or more additional polymeric layers that prevent bacterial escape. Our results show that the biosensor population within LAMPS is able to sense L-lactate production by the mammalian producer cell lines and activate expression of a reporter gene in response. The response of the biosensor cells is proportional to the L-lactate acid concentration and the bacterial whole-cell biosensor does not escape the LAMPS to contaminate the cell culture.

Mapping Cellular Interactions between Cancer and Normal Cells Via Synthetic Notch Ligand/Receptor Pair

- Conference: Mammalian Synthetic Biology Workshop
- Year: 2021
- Proceeding: 2021 Virtual International Mammalian Synthetic Biology Workshop (mSBW)
- Group: General Submissions
- Session:

  Sensors and Multi-input Logic

- Time:

  Monday, July 19, 2021 - 11:45am-12:00pm

Authors:
Avsaroglu, B. - Presenter, UCSF
Roybal, K., University of California, San Francisco
Ibrahim, A., UCSF
Hyrenius-Wittsten, A., University of California San Francisco
Goga, A., University of California San Francisco
Knox, S., University of California San Francisco
Ha, P., University of California San Francisco

Understanding how cancer cells interact within its surroundings is critical for elucidating fundamental mechanisms of cancer biology and metastasis. We utilized the synthetic Notch ligand/receptor pair system (synNotch – Morsut L., Roybal KT., et al. Cell 2016) to track the physical interactions between normal and cancer cells with an ultimate goal of investigating the changes in gene expression profile of normal cells before and after the interactions occur.

Here we use normal epithelial cells (RPE1 hTERT) and adenoid cystic carcinoma (ACC), immortalized patient derived salivary gland cancer cell line). In our study, cell lines express synthetic Notch ligand/receptor pairs to measure the dynamics of touch-based interactions between normal-normal, cancer-cancer and normal-cancer cells. We find that, when co-cultured with ligand-presenting cancer cells, syn-Notch receptor carrying normal cells (1) can internalize the ligand molecules (trans-endocytosis) on average ~500 (+/- 280) times more when compared to receptor negative cells, (2) have detectable ligand molecules in their cytoplasm up to 24 hours post termination of interactions, (3) can be programmed for activation/inactivation of fluorescent reporters for longer term tracking. Moreover, by performing RNA-seq analysis on fluorescently sorted receptor positive normal cells, we compare the gene expression profiles of co-cultured cells with or without trans-endocytosis. Pairwise pathway comparison analysis shows differential regulation of genes with roles in axon signaling, ECM organization, steroid metabolism and Interleukin-10 signaling, providing a basis for physical interaction dependent gene signatures in the cancer microenvironment.

Ultimately, we seek to integrate our system into 3D spheroid/organoid and animal models to identify changes in gene expression levels of non-tumor host cells upon interactions with cancer cells. We are particularly interested with perineural invasion in cancer during the metastatic processes. We envision the same sensor mice will be valuable model to study other processes, such as neurogenesis or development of the immune system.

Temperature-Responsive Optogenetic Probes of Cell Signaling

- Conference: Mammalian Synthetic Biology Workshop
- Year: 2021
- Proceeding: 2021 Virtual International Mammalian Synthetic Biology Workshop (mSBW)
- Group: General Submissions
- Session:

  Sensors and Multi-input Logic

- Time:

  Monday, July 19, 2021 - 12:00pm-12:15pm

Authors:
Bugaj, L. - Presenter, University of Pennsylvania

Optogenetics has provided a powerful toolset to dissect the role of signal dynamics in cells and organisms using light as a precise input. Here we describe a new class of photosensor that is sensitive to both light and temperature. The BcLOV4 photoreceptor is a blue-light sensitive protein that translocates to the membrane and can be used as single-component optogenetic switch to control signaling. We used BcLOV4 to engineer optogenetic control over Ras/Erk and PI3K/Akt signaling, and we found that the probes performed well relative to comparable state-of-the-art optogenetic tools. Surprisingly, we discovered an unanticipated temperature sensitivity of the BcLOV4 protein, such that membrane localization and pathway stimulation spontaneously decay within ~1 hour — despite constant illumination — at a rate proportional to both the temperature and the intensity of light. We systematically characterized this unique dual dependency and we developed a computational model that fully predicts BcLOV4 translocation and signal dynamics across a range of experimental temperatures and light conditions. We also developed new methods to precisely control both light and temperature stimulation of cell cultured in microwell plates. In addition to use in mammalian cells, we show that our BcLOV4-based tools function robustly and stably in both Drosophila and zebrafish without the need for expression-level tuning. Our work provides a new set of easy-to-use optogenetic probes that work across organisms, and a quantitative framework for their optimal use. Moreover, BcLOV4 provides a natural substrate for engineering a new class of tools that respond to light and/or temperature.

Engineering multiple levels of specificity in an RNA viral vector

- Conference: Mammalian Synthetic Biology Workshop
- Year: 2021
- Proceeding: 2021 Virtual International Mammalian Synthetic Biology Workshop (mSBW)
- Group: General Submissions
- Session:

  Foundational Advances

- Time:

  Monday, July 19, 2021 - 1:30pm-2:00pm

Authors:
Gao, X. - Presenter, Caltech

Synthetic molecular circuits could provide powerful therapeutic capabilities, but delivering them to specific cell types and controlling them remains challenging. An ideal "smart" viral delivery system would enable controlled release of viral vectors from "sender" cells, conditional entry into target cells based on cell-surface proteins, conditional replication specifically in target cells based on their intracellular protein content, and an evolutionarily robust system that allows viral elimination with drugs. Here, combining diverse technologies and components, including pseudotyping, engineered bridge proteins, degrons, and proteases, we demonstrate each of these control modes in a model system based on the rabies virus. This work shows how viral and protein engineering can enable delivery systems with multiple levels of control to maximize therapeutic specificity.

Physical Unclonable Functions in Human Cells

- Conference: Mammalian Synthetic Biology Workshop
- Year: 2021
- Proceeding: 2021 Virtual International Mammalian Synthetic Biology Workshop (mSBW)
- Group: General Submissions
- Session:

  Foundational Advances

- Time:

  Monday, July 19, 2021 - 2:00pm-2:30pm

Authors:
Bleris, L. - Presenter, The University of Texas at Dallas

We introduce the genetic Physical Unclonable Functions (PUFs). A PUF is a physical entity which provides a measurable output that can be used as a unique and irreproducible identifier for the artifact wherein it is embedded. Popularized by the electronics industry, silicon PUFs leverage the inherent physical variations of semiconductor manufacturing to establish intrinsic security primitives for attesting integrated circuits. Owing to the stochastic nature of these variations and the multitude of steps involved, photo-lithographically manufactured silicon PUFs are impossible to reproduce (thus unclonable). Inspired by the success of silicon PUFs, we sought to exploit a combination of sequence-restricted barcodes and the inherent stochasticity of CRISPR-induced non-homologous end joining DNA error repair to create the first generation of genetic physical unclonable functions in three distinct human cells (HEK293, HCT116, and HeLa). Genetic PUFs can serve as a foundational principle for establishing provenance attestation protocols for protecting intellectual property and confirming authenticity of engineered cell lines.

High-throughput discovery and characterization of compact transcriptional effectors

- Conference: Mammalian Synthetic Biology Workshop
- Year: 2021
- Proceeding: 2021 Virtual International Mammalian Synthetic Biology Workshop (mSBW)
- Group: General Submissions
- Session:

  Foundational Advances

- Time:

  Monday, July 19, 2021 - 2:30pm-2:45pm

Authors:
Bintu, L. - Presenter, Stanford University

Human gene expression is regulated by thousands of proteins that can activate or repress transcription. To model and manipulate gene expression, we need to measure the strength with which effectors regulate transcription when recruited to a promoter and the dynamics of transcription after they are released (i.e. epigenetic memory). To systematically measure the function of transcriptional effector domains in human cells, we developed a high-throughput assay (HT-recruit) in which pooled libraries of thousands of domains are recruited individually to a reporter gene. Cells are then separated by reporter expression level, and the library of protein domains is sequenced to determine the frequency of each domain in silenced versus active cell populations.

We used this method to:

1) quantify the activation, silencing, and epigenetic memory capability of most of the nuclear protein domains annotated in Pfam, including the KRAB family of over 300 domains. We find that while evolutionary young KRABs are strong repressors, some of the old KRABs are activators.

2) characterize the amino acids responsible for effector function via deep mutational scanning. We applied it to the KRAB domain used in CRISPRi to map the co-repressor binding surface and identify substitutions that improve stability, silencing, and epigenetic memory.

3) discover novel functional domains in unannotated regions of large chromatin regulators, including repressors as short as 10 amino acids.

The compact effectors we have identified so far can be used in synthetic biology and therapeutic applications, where small size is essential for efficient delivery. Our high-throughput method can be easily extended to answer a range of biological questions in processes that already have established fluorescent reporters: DNA repair, RNA splicing, protein degradation, and cell signaling.

Engineered Cells That Record Their Own History

- Conference: Mammalian Synthetic Biology Workshop
- Year: 2021
- Proceeding: 2021 Virtual International Mammalian Synthetic Biology Workshop (mSBW)
- Group: General Submissions
- Session:

  Foundational Advances

- Time:

  Monday, July 19, 2021 - 2:45pm-3:00pm

Authors:
Loveless, T. - Presenter, UC Irvine
Liu, C., University of California At Irvine
Carlson, C., UC Irvine
Liang, G., UC Irvine
Ficht, M., UC Irvine

Mammalian development involves a pattern of disparate but coordinated fate changes undergone by millions to billions of cells over days to years. DNA recording is a technique that has the potential to illuminate these processes at the appropriate scale and depth, by converting transient cellular experiences into durable changes in the genome that can be read out by sequencing. Recently, we published the development of a DNA recorder that writes new DNA, through the repeated insertion of random nucleotides at a single locus in temporal order. Our system, called CHYRON (Cell HistorY Recording by Ordered iNsertion), achieves ordered *in vivo* DNA writing by combining Cas9 nuclease, a self-targeting guide (stg)RNA, and the template-independent DNA polymerase TdT. CHYRON records an average of 14.6 bits per recording site over an average of three rounds, but its recording duration is limited because 25% of edits lead to information-corrupting deletion mutations, which compound over time, *and* the desired insertion mutations increase the length of the stgRNA, which reduces its activity. Now, **we have created a new DNA recording architecture, peCHYRON, that maintains the strengths of CHYRON – ordered, information-dense recording – but (1) drastically reduces the rate of information loss through off-pathway mutations, (2) is in principle compatible with tens of rounds of recording, and (3) can in principle record at least tens of orthogonal inputs.** In each round of peCHYRON recording, prime editor makes one single-strand nick to introduce an insertion mutation that includes both a "signature" sequence that constitutes the record and a "propagation sequence" that enables the next step of recording. In a peCHYRON stable cell line, at least 97% of edits in each round were the desired, information-preserving insertion mutations, and editing efficiency remained constant over at least the first five rounds of recording.

Spatiotemporally Confined Red Light-Controlled Gene Delivery at Single-Cell Resolution Using Adeno-Associated Viral Vectors

- Conference: Mammalian Synthetic Biology Workshop
- Year: 2021
- Proceeding: 2021 Virtual International Mammalian Synthetic Biology Workshop (mSBW)
- Group: General Submissions
- Session:

  Foundational Advances

- Time:

  Monday, July 19, 2021 - 3:00pm-3:15pm

Authors:
Hörner, M. - Presenter, University of Freiburg, CIBSS
Jerez-Longres, C., University of Freiburg
Hudek, A., University of Freiburg
Hook, S., Hannover Medical School
Yousefi, O. S., University of Freiburg
Schamel, W. W. A., University of Freiburg
Hörner, C., Paul-Ehrlich-Institut
Zurbriggen, M., University of Freiburg
Ye, H., East China Normal University
Wagner, H. J., University of Freiburg
Weber, W., SGBM – Spemann Graduate School of Biology and Medicine, University of Freiburg

Methodologies for the controlled delivery of genetic information into target cells are of utmost importance for genetic engineering in both fundamental and applied research. However, available methods for efficient gene transfer into user-selected or even single cells suffer from low throughput, the need for complicated equipment, high invasiveness, or side effects by off-target viral uptake. To overcome these limitations, we engineered an adeno-associated viral (AAV) vector system that transfers genetic information into native target cells upon illumination with cell-compatible red light. This OptoAAV system allows adjustable and spatially resolved gene transfer down to single-cell resolution and is compatible with different cell lines as well as primary cells. Moreover, the sequential application of multiple OptoAAVs enables spatially resolved transduction with different transgenes. The approach presented is likely extendable to other classes of viral vectors and is expected to foster advances in basic and applied genetic research.

Engineering cell fate via cellular reprogramming

- Conference: Mammalian Synthetic Biology Workshop
- Year: 2021
- Proceeding: 2021 Virtual International Mammalian Synthetic Biology Workshop (mSBW)
- Group: General Submissions
- Session:

  Applications in Stem Cells

- Time:

  Tuesday, July 20, 2021 - 10:00am-10:30am

Authors:
Galloway, K. - Presenter, Massachusetts Institute of Technology

Integrating synthetic circuitry into larger transcriptional networks to mediate predictable cellular behaviors remains a challenge within synthetic biology. While significant efforts have been devoted to the design of enhanced synthetic circuitry, less is understood regarding how cellular hardware may impose fundamental performance limitations on integrated circuits. Within the mammalian context, cellular reprogramming continues to generate new cell types, increasingly expanding our perspective of cellular plasticity. Despite improved genetic tools and epigenetic modulations, reprogramming remains a rare cellular event. In this talk, I will describe how I identified molecular roadblocks in reprogramming that arise from tradeoffs between transcription and proliferation rates. Our discovery suggests that topological stress impacts the function of gene networks (e.g. native or synthetic circuits) and constrains cellular transitions. To test this hypothesis, my lab constructs synthetic gene circuits that transmit topological stress into changes in expression to identify plastic cells in cell-fate transitions (e.g. reprogramming, differentiation, cancer). I will discuss how this work and recent findings from my lab open completely new questions about how the structure of the genome stabilizes cellular identity and suggests strategies to improve the design of synthetic gene circuits.

Taking control of stem cell dynamics

- Conference: Mammalian Synthetic Biology Workshop
- Year: 2021
- Proceeding: 2021 Virtual International Mammalian Synthetic Biology Workshop (mSBW)
- Group: General Submissions
- Session:

  Applications in Stem Cells

- Time:

  Tuesday, July 20, 2021 - 10:30am-10:45am

Authors:
Marucci, L. - Presenter, University of Bristol
Pedone, E., University of Bristol
Rocca, D., Bristol Synthetic Biology Centre
Postiglione, L., University of Bristol
di Bernardo, D., Telethon Institute of Genetics and Medicine

Complex non-linear dynamics orchestrate stem cell phenotypes.

Systems biology tools can aid the understanding of origins and mechanisms behind such dynamics but are usually unable to explain if and how they have a role in cell fate determination.

In this talk, I will start discussing combined experimental and computational approaches to model multiscale gene expression, proliferation and spatial dynamics in stem cell cultures, with a focus on multistability of pluripotency genes [1].

I will then propose synthetic biology and applications of control theory to directly engineer gene expression dynamics in stem cells. I will show results obtained using controllers embedded into cells [2], or implemented externally by means of microfluidics/microscopy platforms [3, 4].

I will finally propose the use of cybergenetics methodologies to design superior culture protocols and to automatically prototype mammalian cell non-linear dynamics.

[1] Godwin et al. NPJ Systems Biology and Applications 2017

[2] Pedone et al. Nature Communications 2019

[3] Postiglione et al. ACS Synthetic Biology 2018

[4] Pedone et al. ACS Synthetic Biology 2021

Spatially Controlled Tissue Differentiation Using the Synthetic Receptor Synnotch

- Conference: Mammalian Synthetic Biology Workshop
- Year: 2021
- Proceeding: 2021 Virtual International Mammalian Synthetic Biology Workshop (mSBW)
- Group: General Submissions
- Session:

  Applications in Stem Cells

- Time:

  Tuesday, July 20, 2021 - 10:45am-11:00am

Authors:
Garibyan, M. - Presenter, University of Southern California
Morsut, L., University of Southern California
McCain, M., University of Southern California
Cho, N., University of Southern California
March, A., Children's Hospital LA

Spatial organization of multiple cell types is intrinsic to tissue function in vivo. One goal of tissue engineering is to recapitulate this cellular organization in vitro to create readily available and accurate models for drug screening, disease modeling, and regenerative medicine. However, the field struggles to recreate the organized complexity of native tissues with cellular scale precision. There have been many advancements in the fields of scaffold fabrication and synthetic biology that open new possibilities for tissue engineering. For example, microcontact printing enables researchers to pattern ligands onto a surface with micro-scale precision. A synergistic advancement in synthetic biology has been the Synthetic Notch receptor (synNotch), a modular receptor that allows control of transgene expression upon binding of prescribed ligands. Our goal is to combine these technologies by using microcontact printed ligands to spatially activate the synNotch receptor, and downstream transgenes, in cells. As a case study, we will use this approach to spatially control the differentiation of skeletal muscle cells by activating myoD in fibroblasts, laying the foundation for more accurate, multi-cell tissues in vitro. We have demonstrated that a surface presented ligand, GFP, can activate transgenes in cells via synNotch with spatial control. The activation can be patterned by micro-contact printing the ligand onto a cell adhesive coverslip. The spatial activation of the transgenes leads to spatially controlled myotube formation. Differentiated myotubes immuno-stain positive for alpha-actinin, a prominent myotube marker.

Learning to Program Cellular Machines: Engineering Cells to Treat Disease, Build Tissues, and Elucidate Design Principles

- Conference: Mammalian Synthetic Biology Workshop
- Year: 2021
- Proceeding: 2021 Virtual International Mammalian Synthetic Biology Workshop (mSBW)
- Group: General Submissions
- Session:

  Keynote Speaker: Wendell Lim, University of California, San Francisco

- Time:

  Wednesday, July 21, 2021 - 10:00am-10:45am

Authors:
Lim, W. - Presenter, University of California, San Francisco

The ability to use living cells as a powerful new platform for therapeutics has the potential to transform medicine. With newfound capabilities in genetic engineering and an encyclopedic knowledge of molecular components, we have the potential to engineer cells as smart sense-and-response systems that can be tailored to address the unmet needs of complex tissue-based diseases such as cancer, autoimmunity, and degeneration. Living cells, particularly immune cells, have the unique capability to infiltrate, recognize, coax and kill diseased tissues. Conversely, this cell engineering framework also presents a powerful way to test and systematically understand the general principles by which molecular components are networked to create regulatory circuits capable of precise and robust decision making. In the new science of cellular engineering, many of the most exciting challenges and roadblocks lie in developing general modules for cell sensing, recognition, communication, feedback control and decision making -- modules that could be flexibly used to program diverse cellular behaviors rapidly and predictably. I will describe our efforts to construct such modular cell engineering platforms and to use them to build smarter immune cell therapies, as well as to build self-organizing multi-cellular systems.

Clinically-Driven Design of Synthetic Gene Regulatory Programs in Therapeutic Human Cells

- Conference: Mammalian Synthetic Biology Workshop
- Year: 2021
- Proceeding: 2021 Virtual International Mammalian Synthetic Biology Workshop (mSBW)
- Group: General Submissions
- Session:

  Immunotherapy and Cell Therapy

- Time:

  Wednesday, July 21, 2021 - 10:45am-11:15am

Authors:
Khalil, A. S. - Presenter, Harvard University

Synthetic biology seeks to enable the rational design of regulatory circuits to reprogram and control cellular function. The application of this approach to human cells could lead to powerful gene and cell-based therapies that provide transformative ways to combat complex diseases. Realizing this vision will require engineering synthetic tools and systems to meet clinical requirements. In this talk, I will describe our clinically-driven design process to create synthetic toolkits for custom gene regulation in therapeutic human cells. Specifically, we have developed synthetic transcription regulators (SynZiFTRs) that feature a compact human protein-based design, enable precise genome-orthogonal regulation, and can be modulated by FDA-approved small molecules. Using SynZiFTRs, we engineer human primary T cells with genetic programs that enable external, drug-regulated control over immunotherapeutic genes and therapeutic cellular activity *in vivo*. SynZiFTRs can also be used in the context of compact, fully humanized, and customizable synthetic receptors (termed SNIPRs) for autonomous control of therapeutic T cells. These developments establish synthetic platforms for custom gene regulation in human cells with the potential to accelerate clinical translation of synthetic systems.

Engineering Next-Generation T Cells for Cancer Immunotherapy

- Conference: Mammalian Synthetic Biology Workshop
- Year: 2021
- Proceeding: 2021 Virtual International Mammalian Synthetic Biology Workshop (mSBW)
- Group: General Submissions
- Session:

  Immunotherapy and Cell Therapy

- Time:

  Wednesday, July 21, 2021 - 11:15am-11:45am

Authors:
Chen, Y. Y. - Presenter, University of California, Los Angeles

The adoptive transfer of T cells expressing chimeric antigen receptors (CARs) has demonstrated clinical efficacy in the treatment of advanced cancers, with anti-CD19 CAR-T cells achieving up to 90% complete remission among patients with relapsed B-cell malignancies. However, challenges such as antigen escape and immunosuppression limit the long-term efficacy of adoptive T-cell therapy. Here, I will discuss the development of and clinical data on next-generation T cells that can target multiple cancer antigens and resist antigen escape. I will also present recent work on tuning CAR signaling activities via rational protein design to achieve greater *in vivo* anti-tumor efficacy. This presentation will highlight the potential of synthetic biology in generating novel mammalian cell systems with multifunctional outputs for therapeutic applications.

A High-Throughput Exploration of the CAR-T Signaling Landscape Identifies Novel Domains for Next-Generation Immunotherapies

- Conference: Mammalian Synthetic Biology Workshop
- Year: 2021
- Proceeding: 2021 Virtual International Mammalian Synthetic Biology Workshop (mSBW)
- Group: General Submissions
- Session:

  Immunotherapy and Cell Therapy

- Time:

  Wednesday, July 21, 2021 - 11:45am-12:00pm

Authors:
Goodman, D. - Presenter, University of California, San Fransisco
Roybal, K., University of California, San Francisco
Azimi, C., University of California, San Francisco
Marson, A., University of California, San Francisco
Kearns, K., University of California, San Diego
Garakani, K., University of California, San Francisco
Bluestone, J. A., Diabetes Center and the Sean N. Parker Autoimmunity Laboratory

Chimeric Antigen Receptors (CARs) are the first successful synthetic immune receptors for adoptive T cell therapy. CARs with different signaling domains (such as CD28 and 4-1BB) can modulate distinct T-cell functions including cytokine secretion, cytotoxicity, and overall persistence in the patient. Identification and optimization of new CAR signaling domains will help to extend their therapeutic applications to solid tumors and indications beyond cancer. Here we use a pooled screening approach which combines fluorescence-activated cell sorting and high-throughput sequencing to conduct a broad search of the CAR signaling landscape in primary human T cells. We measured CARs containing 40 different domains - including those not normally associated with T-cell costimulation - for activation, cytokine secretion, and proliferation in response to multiple weeks of repetitive antigen challenge. These domains differed in activity between CD4 and CD8 T cells and between early and late proliferators. We identified and further characterized several potent costimulatory domains beyond 4-1BB and CD28 which enhance proliferation (CD40) and cytotoxicity (BAFF-R and TACI) after repeated antigen exposure *in vitro* and *in vivo*. We also identified a potent co-inhibitory domain, KLRG1, which strongly attenuates multiple aspects of T cell function. Via single-cell RNA and surface protein analysis, we show that these domains are enriched in distinct functional states which differ in markers of T cell memory, metabolism, and innateness. In particular, CAR-T cells containing the BAFF-R costimulatory domain, normally associated with B-cell survival, exhibit improved cytotoxicity, proliferative capacity, and a functional program which is correlated with positive clinical outcomes in multiple studies. These new domains could enhance CAR clinical efficacy and their identification shows how mapping the costimulatory signaling landscape will enhance our ability to engineer diverse functions into therapeutic primary human T cells.

Directing Cellular Interactions and Organization with Synthetic Adhesion Receptors

- Conference: Mammalian Synthetic Biology Workshop
- Year: 2021
- Proceeding: 2021 Virtual International Mammalian Synthetic Biology Workshop (mSBW)
- Group: General Submissions
- Session:

  Immunotherapy and Cell Therapy

- Time:

  Wednesday, July 21, 2021 - 12:00pm-12:15pm

Authors:
Stevens, A. - Presenter, University of California-San Francisco
Lim, W., University of California, San Francisco
Gerdts, J., University of California-San Francisco
Fletcher, D. A., University of California, Berkeley
Harris, A., University of California, Berkeley

Cellular adhesion molecules have evolved to control an array of intracellular signaling functionalities that are critical to the regulation of multicellular organisms. In order to design artificial multicellular systems capable of complex organization and behavior, we have engineered a toolkit of synthetic cellular adhesion molecules with programmable extracellular recognition domains that activate endogenous signaling. This engineering strategy enables control of both the manner of extracellular engagement (e.g. homophilic or heterophilic) and choice of intracellular signaling program. For example, some of these synthetic receptors mediate tight cell-cell interactions via cytoskeletal reorganization, while others facilitate receptor enrichment through the recruitment of scaffold proteins. Using these receptors, we explore the interplay between extracellular affinity and intracellular signaling output in dictating cellular adhesion strength. Moreover, we examine how introducing asymmetric signaling interfaces not present in nature impacts cell-cell adhesion. We then apply this system to form multicellular structures with precise cellular architecture and defined compartmentalization. Lastly, we integrate synthetic adhesion to modulate native cell interactions from endogenous adhesion molecules. We are presently employing this new capability to investigate the relationship between cellular patterning and tissue function and to design T cells capable of custom tissue trafficking.

Synthetic Multistability in Mammalian Cells

- Conference: Mammalian Synthetic Biology Workshop
- Year: 2021
- Proceeding: 2021 Virtual International Mammalian Synthetic Biology Workshop (mSBW)
- Group: General Submissions
- Session:

  Gene Regulation

- Time:

  Wednesday, July 21, 2021 - 1:45pm-2:00pm

Authors:
Zhu, R. - Presenter, California Institute of Technology
Elowitz, M., California Institute of Technology
Garcia-Ojalvo, J., Universitat Pompeu Fabra
del Rio-Salgado, J. M., Tecnologico de Monterrey

In multicellular organisms, gene regulatory circuits generate thousands of molecularly distinct, mitotically heritable states, through the property of multistability. Designing synthetic multistable circuits would provide insight into natural cell fate control circuit architectures and allow engineering of multicellular programs that require interactions among cells in distinct states. Here we introduce MultiFate, a naturally-inspired, synthetic circuit that supports long-term, controllable, and expandable multistability in mammalian cells. MultiFate uses engineered zinc finger transcription factors that transcriptionally self-activate as homodimers and mutually inhibit one another through heterodimerization. Using model-based design, we engineered MultiFate circuits that generate up to seven states, each stable for at least 18 days. MultiFate permits controlled state-switching and modulation of state stability through external inputs, and can be easily expanded with additional transcription factors. Together, these results provide a foundation for engineering multicellular behaviors in mammalian cells.

Shield: A Platform for Fast and Simple Screening of Anti-Silencing DNA Elements in Human Cells

- Conference: Mammalian Synthetic Biology Workshop
- Year: 2021
- Proceeding: 2021 Virtual International Mammalian Synthetic Biology Workshop (mSBW)
- Group: General Submissions
- Session:

  Gene Regulation

- Time:

  Wednesday, July 21, 2021 - 2:00pm-2:15pm

Authors:
Zhang, M. - Presenter, University of Illinois At Urbana-Champaign
Zhao, H., University of Illinois-Urbana

Transgene expression in mammalian cells plays an essential role in both fundamental research and the biotechnology industry. However, stably integrated transgene is subject to epigenetic silencing, which represents a major bottleneck in this field. Numerous regulatory elements (e.g., cHS4, UCOE, S/MARs) have been adopted to overcome transgene silencing, yet they are associated with various limitations such as reduced viral titer and context-dependent activity. Therefore, there are continuing efforts to discover novel anti-silencing elements with better functionalities. Here we report SHIELD (**S**ite-specific **H**eterochromatin **I**nsertion of DNA **E**lements at **L**amina-associated **D**omains), a platform for fast and simple screening of anti-silencing DNA elements in human cells. SHIELD combines two distinct genome engineering technologies together (CRISPR and serine integrase) to achieve targeted insertion of large DNA fragments (5-10kb) at highly compact heterochromatin loci with high efficiency. Compared with the classical barrier assay which takes up to 40~60 days to fully observe the silencing effect, SHIELD expedites the process by placing the reporter cassette at a highly repressive heterochromatin locus, thus significantly shortening the time frame to ~14 days. Moreover, due to the high efficiency and single-copy insertion feature of SHIELD, polyclonal cells can be directly used for downstream analysis, circumventing the need for clonal isolation or the control of integrant copy number. Using SHIELD, we demonstrated the screening of >20 candidates from *in silico* prediction and identified 3 elements with activity equal or higher than the positive control. Taken together, we anticipate the SHIELD platform would accelerate the discovery of novel anti-silencing DNA elements as well as the functional assessment of engineered variants in human cells.

Rewiring a Calcium-Dependent Signaling Pathway in Mammalian Cells

- Conference: Mammalian Synthetic Biology Workshop
- Year: 2021
- Proceeding: 2021 Virtual International Mammalian Synthetic Biology Workshop (mSBW)
- Group: Poster Session
- Session:

  Poster Session

Authors:
Bencina, M. - Presenter, National Institute of Chemistry
Meško, M., National Institute of Chemistry
Lebar, T., National Institute of Chemistry
Dekleva, P., National Institute of Chemistry
Jerala, R., National Institute of Chemistry

Methods for controlling and regulating cellular processes are of key importance in the basic research of cell biology and understanding of cellular mechanisms. Synthetic systems for regulation of gene expression are often developed based on known cell signaling pathways. Calcium-dependent regulation is an attractive approach for achieving such control, as diverse stimuli induce calcium influx by activating membrane receptors. Here, we employed engineered NFAT transcription factors to induce the potent and robust activation of exogenous gene expression in mammalin cells. Furthermore, we designed a calcium-dependent transcription factor that does not interfere with NFAT-regulated promoters and potently activates transcription in several mammalian cell types. Additionally, we demonstrate that coupling the circuit to a calcium-selective ion channel resulted in capsaicin- and temperature-controlled gene expression. This engineered calcium-dependent signaling pathway enables tightly controlled regulation of gene expression through different stimuli in mammalian cells and is versatile, adaptable, and useful for a wide range of therapeutic and diagnostic applications.

A Perturbation Approach to Quantitatively Decipher Transcription Factor Kinetic Response to Target Gene Output In Vivo

- Conference: Mammalian Synthetic Biology Workshop
- Year: 2021
- Proceeding: 2021 Virtual International Mammalian Synthetic Biology Workshop (mSBW)
- Group: Poster Session
- Session:

  Poster Session

Authors:
Singh, A. - Presenter, Princeton University

In early metazoan development, cells must decide their fate to create robust spatial gene expression patterns leading to proper body segmentation. In the fruit fly, these early processes are tightly regulated by maternal and zygotic transcription factors (TFs). One of the classical maternal TF, Bicoid (Bcd), is known to organize spatial and temporal patterns of target genes called gap genes. Previous attempts by genetic and optogenetic approaches had limited temporal control to distinguish how TFs directly or indirectly regulate gap gene expression in space and time. Here we present a light-inducible system to regulate Bcd TF concentration in living fly embryos on second time-scale. Precise temporal optogenetic perturbation allows us to quantify the response time of gene activation to input factor perturbation in living embryos. We show how nuclear concentration and TF residence times affect the kinetic rates and expression levels of downstream gene targets. Furthermore, we explore how cells utilize differential levels of TF to make cell fate decisions. Our synthetic approach to modulate TF concentration while simultaneously recording transcriptional gene output in real-time presents a powerful tool to study transcriptional dynamics during development.

Development of novel RNA-based regulators for mammalian synthetic biology

- Conference: Mammalian Synthetic Biology Workshop
- Year: 2021
- Proceeding: 2021 Virtual International Mammalian Synthetic Biology Workshop (mSBW)
- Group: Poster Session
- Session:

  Poster Session

Authors:
Mol, A. - Presenter, University of Edinburgh
Rosser, S. J., University of Edinburgh

Mammalian regulatory devices use diverse mechanisms to allow flexible, precise, and comprehensive control over gene expression and cellular development. Fine-tuning of gene expression is critical for many synthetic biology applications. More recently, RNA-based regulatory devices have been explored for modulating mammalian gene expression. Due to their structural properties, riboregulators provide a convenient basis for the development of ligand-dependent controllable systems. In mammalian systems, insertion of self-cleaving ribozymes in the 5' or 3' untranslated region of the mRNA of the gene have been utilized for controlling gene expression on a post-transcriptional level. Ribozyme cleavage of the respective mRNA results in RNA degradation and hence significant downregulation of gene expression, whereas ribozyme inhibition upregulates the encoded protein expression. In this project, we are focused to develop novel aptazyme switches for the control of gene regulation with potential application in e.g. construction of mammalian synthetic circuits and gene therapy vectors. Aptazymes can be engineered by attaching ligand-sensing aptamer domain to the catalytically active ribozyme. Ligand-responsive ribozyme switches can regulate self-cleavage events in mRNAs to modulate the stability of the transcript in response to ligand levels. The utilization of aptazymes in synthetic biology and therapeutic applications has many advantages such as an increased robustness due to in cis regulation, small coding space and a high degree of modularity.

Continuous Ultrasound Imaging of Second-Generation Mammalian Acoustic Reporter Genes Expressed Using a Novel Ratio-Tunable Mammalian Polycistronic Expression System.

- Conference: Mammalian Synthetic Biology Workshop
- Year: 2021
- Proceeding: 2021 Virtual International Mammalian Synthetic Biology Workshop (mSBW)
- Group: Poster Session
- Session:

  Poster Session

Authors:
Duan, M. - Presenter, California Institute of Technology
Shapiro, M. G., California Institute of Technology
Farhadi, A., California Institute of Technology
Hurt, R. C., Caltech
Dutka, P., California Institute of Technology
You, M. Y., Caltech
Swift, M., California Institute of Technology

Ultrasound is an established technology for noninvasive deep-tissue imaging that overcomes the low penetration depth of optical methods. However, until recently it was impossible to couple ultrasound contrast to gene expression or cellular function. Gas vesicles (GVs) were recently introduced as the first genetically encoded ultrasound contrast agents expressed from acoustic reporter genes (ARGs) in bacteria and mammalian cells. However, the first-generation mammalian acoustic reporter genes (mARGs), based on 9 essential GV genes from a soil bacterium *B. megaterium*, suffer from a number of limitations: linking individual GV genes into compact mammalian polycistronic expression cassettes severely attenuates GV expression and ultrasound contrast compared to the cotransfection of individual genes, both transient and stably transduced mARGs require coexpression of three separate polycistronic cassettes for GV production, and the acoustic properties of *B. megaterium* GVs only allow imaging using a destructive ultrasound modality, limiting their use to endpoint measurements. Here we present a new generation of mARGs that overcome the above-mentioned limitations. We demonstrate that mARGs2.0 provide strong expression from only two polycistronic expression cassettes with optimized cassette ratio, nonlinear acoustic properties that allow for nondestructive continuous imaging using amplitude modulation pulse sequences, and the ability to modulate acoustic properties based on gene composition. Additionally, we are developing a novel ratio-tunable mammalian polycistronic expression system that enables expression of mARGs2.0 from single transcripts. We extensively characterize mARGs2.0 *in vitro* and demonstrate their utility *in vivo* by imaging gene expression in an orthotopic cancer model. For this purpose, we stably integrated these reporter genes into the genomes of human breast cancer cells and imaged them nondestructively *in vivo* after tumor formation in mouse mammary fat pads and chemically induced expression.

Mutational Analysis of a Transgenic Mouse Model of Human Sry

- Conference: Mammalian Synthetic Biology Workshop
- Year: 2021
- Proceeding: 2021 Virtual International Mammalian Synthetic Biology Workshop (mSBW)
- Group: Poster Session
- Session:

  Poster Session

Authors:
Pelosi, E. - Presenter, University Of Queensland
Thomson, E., University of Queensland
Zhao, L., University of Queensland
Chen, Y. S., Indiana University School of Medicine
Sreenivasan, R., Murdoch Children's Research Institute
Croft, B., 4. Murdoch Children's Research Institute
Sinclair, A., Murdoch Children's Research Institute
Weiss, M., Indiana University School of Medicine
Koopman, P., 1. Institute for Molecular Bioscience

Testis development in males is controlled in most eutherian mammals, including humans, by the Y-chromosome gene SRY. Mutations of SRY result in differences/disorders of sex development (DSD), congenital conditions characterized by atypical development of the reproductive system. Thirty decades after its discovery, we still have an incomplete understanding of how this gene operates, and our translational efforts are dramatically hindered by the structural differences between mouse and human SRY. Most importantly, the lack of a system to experimentally generate mutations of human SRY and study their effect in vivo has limited progress in defining the mechanisms underlying the role of SRY in DSD. We have addressed this issue by generating a mouse model carrying a human SRY transgene able to drive male sex determination in XX mice. Using CRISPR-Cas9, we conducted a structure-function analysis of SRY by introducing novel modifications in vivo in each of its three domains (N-terminal, HMG box, and C-terminal), and characterizing the properties of the resulting variants. We showed that specific modifications resulted in complete or partial loss of SRY function, and that variation of the C-terminal - but not HMG box - domain affects protein stability rather than transcriptional activation capacity. Our results provide new functional insights unique to human SRY and the causes of DSD and present a versatile and powerful system for the study of SRY variations. This unique strategy has broad application potential, including in vivo functional studies of SRY variations, queries of the SRY mutation database, and the characterization of new pathogenic SRY variations found in clinical DSD.

Lentivirus Production and Transduction of T-Cells Using Packaging Cells Engineered to Secrete Active Nuclease into Serum-Free Media

- Conference: Mammalian Synthetic Biology Workshop
- Year: 2021
- Proceeding: 2021 Virtual International Mammalian Synthetic Biology Workshop (mSBW)
- Group: Poster Session
- Session:

  Poster Session

Authors:
Ali, S. - Presenter

To improve the purity of lentiviral vectors (LVV) for cell and gene therapy applications manufacturersâ€™ currently remove residual DNA using a DNase treatment step. This is a prohibitively expensive but necessary step in the manufacturing process as regulators require concentrations of less than 10 ng/dose at sizes less than 200 base pairs in the final product. As an alternative we have engineered serum-free adapted HEK293T cells to express nuclease enzymes to remove DNA impurities from LVV supernatant. Three nuclease expression cassettes were designed by fusing the N-terminus of the *Staphylococcus aureus* nuclease B (nucB) open reading frame to: the native *S. aureus* nuclease signal peptide, the mammalian murine IgÎº chain leader sequence, and a virus-encoded transport protein, under the Tet repressor system. All three nuclease expression cassettes demonstrated levels of activity equivalent to or better than Benzonase at 250 units in growth medium harvested from engineered cell lines 24 hours post tetracycline induction. When treating 1.5 µg 1 kb DNA ladder (500 bp to 10 Kbp), incubations at 37°C for 2-hours with 10 µL of growth media reduced DNA ladder to non-visible sizes on 1% agarose gel. To verify that a nuclease secreting cell line can support LVV production, a LVV encoding for green fluorescent protein was prepared by transient transfection. Nuclease activity was demonstrated in fractions collected during LVV production and had no measurable effect on infectious titer of the clarified supernatant when transducing both HEK293T and AGF-T cells under serum-free conditions. This provides a holistic approach to remove residual DNA early in the process stream and improve the purity of the final product, whilst avoiding the addition of exogenous nuclease and its associated costs at scale. This may also potentially reduce any deficiencies of downstream processing attributed to the viscosity of residual DNA.

Interfacing Electrochemistry with Synthetic Gene Circuits

- Conference: Mammalian Synthetic Biology Workshop
- Year: 2021
- Proceeding: 2021 Virtual International Mammalian Synthetic Biology Workshop (mSBW)
- Group: Poster Session
- Session:

  Poster Session

Authors:
Dai, Y. - Presenter, Duke University
Caplan, A. I., Case Western Reserve University
Liu, C. C., Case Western Reserve University

Synthetic biology enables modular engineering and arrangement of functional nucleic acids and proteins to act cooperatively in response to a biorecognition event by initiating logic dictated specific biochemical reaction circuit. In other words, orthogonal biomolecules, implemented as LEGO bricks, are arranged into a cascade to deliver sequence encoded functions. Such engineering approaches are on-track to introduce a new paradigm for modern medicine and biotechnology, such as gene-circuit based chassis for regenerative medicine, nucleic acid computers capable of performing complex logic, gene-circuit based biosensors able to detect broad ranges of biomolecules. Specifically, highlighted by the current COVID-19 pandemic, cell-free synthetic gene circuit has significantly promoted the development of biosensing systems with high specificity, high sensitivity, and low interference, bringing in a powerful design approach for point-of-care system. However, synthetic biology only leverages the information processing capability of biological machineries to tackle molecular recognition and cascading reactions. Comparing with electrical system, synthetic gene circuit does not have the capability to extract processed information for display or memory storage, emphasizing the necessity of a physical interface to curate and translate the downstream biomolecular output into a readable and digital processable signal. Herein, I will discuss engineering strategies to interface electrochemistry with synthetic biology for the development of universal biosensing systems capable of detecting nucleic acids and proteins, translating biomolecular information directly into electrical current. I will first discuss our efforts in designing and developing the first electrochemistry based CRISPR biosensor (E-CRISPR), in which we constituted an electrochemical surface to probe the trans-cleavage activity of Cas12a. I will further introduce the design of a multi-function heterogeneous biochemical circuit and the utilization of electrochemistry to probe the circuit output. Overall, these two examples demonstrate the robust capability of the unique combination of electrochemistry and synthetic biology for the development of next-generation bioanalytical system.

Robust and Tunable Signal Processing in Mammalian Cells Via Engineered Covalent Modification Cycles

- Conference: Mammalian Synthetic Biology Workshop
- Year: 2021
- Proceeding: 2021 Virtual International Mammalian Synthetic Biology Workshop (mSBW)
- Group: Poster Session
- Session:

  Poster Session

Authors:
Jones, R. - Presenter, MIT
Qian, Y., MIT
Ilia, K., MIT
Wang, B., MIT
Laub, M., MIT
Del Vecchio, D., MIT
Weiss, R., Massachusetts Institute of Technology

Rewired and synthetic signaling networks can impart cells with new functionalities and enable efforts in engineering cell therapies and directing cell development. However, there is a need for tools to build synthetic signaling networks that are tunable, can precisely regulate target gene expression, and are robust to perturbations within the complex context of mammalian cells. Here, we use proteins derived from bacterial two-component signaling pathways to develop synthetic phosphorylation-based devices for tunable, precise, and robust gene regulation in mammalian cells. First, we isolate kinase and phosphatase proteins from the bifunctional histidine kinase EnvZ. We then use these proteins to engineer a synthetic covalent modification cycle, in which the kinase and phosphatase competitively regulate phosphorylation of the cognate response regulator OmpR, thereby tuning OmpR-driven gene expression. Further, we show that this tunability can be extended by connecting phosphatase expression to small molecule and miRNA inputs in the cell, with the latter enabling cell-type specific signaling responses and accurate cell type classification. Finally, we implement a tunable negative feedback controller by co-expressing the kinase-driven output gene with the small molecule-tunable phosphatase. This negative feedback substantially reduces cell-to-cell noise in output expression and mitigates the effects of cell context perturbations due to off-target regulation and resource competition. Our work thus lays the foundation for establishing tunable, precise, and robust control over cell behavior with synthetic signaling networks.

Activation of Synthetic Notch Gene Networks Via Material-Conjugated Synthetic Ligands

- Conference: Mammalian Synthetic Biology Workshop
- Year: 2021
- Proceeding: 2021 Virtual International Mammalian Synthetic Biology Workshop (mSBW)
- Group: Poster Session
- Session:

  Poster Session

Authors:
Hoffman, T. - Presenter, UCLA
Khademhosseini, A., University of California-Los Angeles (UCLA)
Morsut, L., University of Southern California
Li, S., UCLA
Garibyan, M., University of Southern California
Kim, H., University of California Los Angeles

Synthetic notch (synNotch) receptors have emerged as powerful tools in controlling cell differentiation and morphogenesis[1]. SynNotch user-defined networks have great potential for robustly controlling cell behavior by enabling environmental ligands to directly (no signaling intermediates) induce target gene expression. This approach can improve the multiplexed and spatial control of cell phenotype for the development of more complex 3D multicellular tissues, improving the relevancy of tissue models and therapeutics. However, synNotch activation has relied on cell-cell contact with sender cells that present membrane-tethered ligands. Cell-presented ligand systems require an additional cell engineering step and remain challenging to control in 3D environments. To enable the control over synNotch gene networks in engineered 3D tissues, the development of novel tools that can both present ligands and retain the capacity to activate mechanosensitive synNotch receptors are required.

To this aim, we have demonstrated that the covalent immobilization of synthetic ligands to materials can efficiently activate synNotch gene networks. Green fluorescent protein (GFP) was conjugated to microparticles and hydrogel polymers by carbodiimide or click chemistry linkers, respectively. Murine fibroblasts were engineered to express GFP-responsive synNotch receptors that drive the expression of reporter protein, mCherry. SynNotch engineered cells were cocultured with GFP-conjugated microparticles or encapsulated within GFP-conjugated gelatin methacryloyl hydrogels, and the extent of synNotch activation was quantified based on the percent and intensity of mCherry-expressing cells. We have found that material-conjugated ligands can induce synNotch gene activation similar to ligand-presenting sender cells. Modular conjugation approaches and the capacity for multi-ligand presentation can be used as a platform to spatially control target gene activation in 3D hydrogels. In future work, we will characterize the effect of material properties on SynNotch activation and use these tools to control multi-fate differentiation in 3D engineered tissues.

1. Morsut L, et al. Cell. 2016;164(4):780-91.

Precise Tumor Targeting with NOT Logic-Gated Chimeric Antigen Receptor Gene Circuits

- Conference: Mammalian Synthetic Biology Workshop
- Year: 2021
- Proceeding: 2021 Virtual International Mammalian Synthetic Biology Workshop (mSBW)
- Group: Poster Session
- Session:

  Poster Session

Authors:
Lee, S. - Presenter, Boston University
Frankel, N., Senti Biosciences
Lee, D., Senti Biosciences
Gainer, M., Senti Biosciences
Wood, T., Senti Biosciences
Gorman, W., Senti Biosciences
Gordley, R., Senti Biosciences
Guzman-Ayala, M., Senti Biosciences
Lu, T., Senti Biosciences
Lee, G., Senti Biosciences
Wong, W., Boston University
Garrison, B., Senti Biosciences

Immune cells engineered with chimeric antigen receptors (CARs) have produced breakthrough efficacy in cancer immunotherapy. Still, the lack of uniquely defined cancer-specific antigens brings risk of life-threatening on-target, off-tumor toxicity. CAR suppression in response to a "safety antigen" selectively expressed on healthy cells could broaden the spectrum of addressable cancers and enable more aggressive on-target, on-tumor treatments. Here, we engineered NK and T cells with a synthetic NOT GATE circuit: an activating CAR (aCAR) killed target cells expressing activating antigen, while an inhibitory CAR (iCAR) suppressed cytotoxicity against target cells also expressing a safety antigen.

We have identified a panel of functional iCARs with different intracellular domains in NK cells and T cells. NOT GATE-equipped CAR-NK or -T cells pairing an iCAR with a second-generation aCAR significantly reduced cytotoxicity and secretion of inflammatory cytokines in response to safety antigen-expressing target cells, while aCAR-only expressing cells exhibited no such significant differences. The NOT GATE gene circuit did not interfere with aCAR performance in the absence of the safety antigen. We demonstrated the persistence of iCAR functionality with repetitive exposure of engineered immune cells to the target cells.

We have developed a NOT GATE gene circuit and validated its function in CAR-NK and -T cells. NK cells can be used in an allogeneic fashion, which may lead to a reduced cost of therapy and increased patient access relative to autologous products. Immune cell therapies equipped with our CAR NOT GATE gene circuits may improve safety for currently validated cancer antigens and enable new treatments for previously unaddressed indications.

Sry Length Regulates Its Cellular Stability and Hence The Robustness of Testis Determination

- Conference: Mammalian Synthetic Biology Workshop
- Year: 2021
- Proceeding: 2021 Virtual International Mammalian Synthetic Biology Workshop (mSBW)
- Group: Poster Session
- Session:

  Poster Session

Authors:
Chen, Y. S. - Presenter, Indiana University School of Medicine
Thomson, E., University of Queensland
Koopman, P., 1. Institute for Molecular Bioscience
Pelosi, E., University Of Queensland
Weiss, M., Indiana University School of Medicine

A seeming paradox is posed by metazoan gene-regulatory networks (GRNs) that are robust yet evolvable. Insight is obtained through studies of bistable genetic circuits mediating developmental decisions and in particular transcriptional thresholds operated by a key transcription factor (TF). A model is provided by human SRY, a Y-encoded TF that initiates testicular differentiation, in turn enabling male development.

Function domains in SRY (204 residues) cluster in its high mobility group (HMG) box (residues 56-141). The functions of N- and C-terminal non-box segments are not well characterized. In addition, the paucity of experimental systems connecting SRY structure to function (as tested in cell models and *in vivo*) has limited progress. Here, we have sought to measure the transcriptional threshold of SRY based on cell-based and transgenic studies of a genetic boundary between organization and dysgenesis. Our findings demonstrate a threshold length in the C-terminal domain of human SRY that determines the protein's proteasome-enforced half-life in the cell. In a rodent pre-Sertoli cell line, the consequent reduction in the intracellular concentration of truncated SRY constructs is associated with a twofold attenuation of the male-specific downstream GRN. *In vivo* CRISPR-Cas9-edited XX mice expressing the 1-164 fragment of human SRY cannot redirect to male differentiation whereas that expressing 1-200 of SRY exhibit male GRN development.

This study provides insight into the robustness of human SRY and illustrates a powerful strategy to link biochemical properties in cell culture to developmental outcomes *in vivo*. Our study unmasks a checkpoint in a mammalian TF serving as a key node in a sex-specific GRN. An experimental "control knob" in cell-based studies is provided by proteasome inhibition. Our approach probes molecular determinants of cell fate and so promises to extend structure-function studies of SRY from its HMG box to flanking non-box domains.

Rapid Directed Molecular Evolution of Fluorescent Proteins in Mammalian Cells

- Conference: Mammalian Synthetic Biology Workshop
- Year: 2021
- Proceeding: 2021 Virtual International Mammalian Synthetic Biology Workshop (mSBW)
- Group: Poster Session
- Session:

  Poster Session

Authors:
Babakhanova, S. - Presenter, Stanford
Jung, E., University of Illinois at Chicago
Namikawa, K., Technische Universität Braunschweig
Subach, O., National Research Center "Kurchatov Institute"
Korzhenevskiy, D., National Research Center "Kurchatov Institute"
Rakitina, T., National Research Center "Kurchatov Institute"
Xiao, X., Westlake University
Cosio, D., Media Lab, Massachusetts Institute of Technology
Shi, J., Westlake Laboratory of Life Sciences and Biomedicine, Westlake University
Wu, S., Westlake Laboratory of Life Sciences and Biomedicine, Westlake University
Park, D., Media Lab, Massachusetts Institute of Technology
Eisenhard, L., Division of Cellular and Molecular Neurobiology, Zoological Institute, Technische Universität Braunschweig
Tang, H., Westlake Laboratory of Life Sciences and Biomedicine, Westlake University
Sun, Y., Westlake Laboratory of Life Sciences and Biomedicine, Westlake University
Köster, R., Division of Cellular and Molecular Neurobiology, Zoological Institute, Technische Universität Braunschweig
Subach, F., National Research Center "Kurchatov Institute"
Piatkevich, K., Westlake Laboratory of Life Sciences and Biomedicine
Boyden, E. S., Massachusetts Institute of Technology

In vivo imaging of model organisms is heavily reliant on fluorescent proteins with high brightness. Here we describe a practical method for rapid optimization of fluorescent proteins via directed molecular evolution in mammalian cells. We employed this approach to enhance intracellular brightness of a set of the spectrally diverse fluorescent proteins. The developed proteins demonstrated high performance for fluorescent labeling of neurons in culture and *in vivo* in model organisms such as *C.elegans*, *Drosophila* fruit flies, and zebrafish. The described method has great potential to be adopted by protein engineers due to its simplicity and convenience. We also believe that the new enhanced FPs will find wide application for *in vivo* imaging of small model organisms.

Games: An ODE Model Development Workflow for Systematic Characterization of Synthetic Genetic Parts

- Conference: Mammalian Synthetic Biology Workshop
- Year: 2021
- Proceeding: 2021 Virtual International Mammalian Synthetic Biology Workshop (mSBW)
- Group: Poster Session
- Session:

  Poster Session

Authors:
Dray, K. - Presenter, Northwestern University
Muldoon, J., UCSF
Mangan, N. M., University of Washington
Bagheri, N., University of Washington
Leonard, J. N., Northwestern University

Mathematical modeling is invaluable for advancing our understanding and design of synthetic biological systems. However, the model development process is complicated and often unintuitive, requiring iteration on various computational tasks and comparisons with experimental data. Moreover, ad hoc model development can lead to models that lack objectivity and reproducibility. To help practitioners manage these challenges, we introduce **GAMES**: a workflow for **G**eneration and **A**nalysis of **M**odels for **E**xploring **S**ynthetic systems that includes a combination of both automated and human-in-the-loop processes. We systematically consider the process of developing models, including model formulation, parameter estimation, parameter identifiability analysis, experimental design, model reduction, model refinement, and model selection. We demonstrate the workflow with a case study on a chemically responsive transcription factor. The generalizable workflow presented here can enable biologists to more readily build and analyze models for a variety of applications.

Chimeric Antigen Ligand (CAL) T Cell Therapy for the Prevention of Allograft Rejection

- Conference: Mammalian Synthetic Biology Workshop
- Year: 2021
- Proceeding: 2021 Virtual International Mammalian Synthetic Biology Workshop (mSBW)
- Group: Poster Session
- Session:

  Poster Session

Authors:
Siddiqui, M. - Presenter, Boston University

There are nearly 2 million people living with limb loss in the US, where over 185,000 amputations occur each year. Over the last decade, our capabilities to conduct reconstructive transplantation, such as vascularized composite allotransplantation (VCA) from diseased donors, has vastly improved to become a viable alternative for some candidates. However, donors and recipients of these transplants are commonly not HLA identical, resulting in an immune response due to interactions between the recipient's alloreactive T cells that recognize "non-self" antigens and the donor cells that present these mismatched "non-self" MHC-antigen complexes. The most common method of inducing tolerance to the donor organ is by lifelong administration of immunosuppressive drugs that come with the risk of opportunistic infections and toxicities. As such, there is an urgent need for a highly specific cell targeting technology to restore immune tolerance without the need for life-long immunosuppression. Our approach to achieving this is to engineer split Chimeric Antigen Ligand (CAL) T cells that use TCR-targeting adapter molecules to specifically deplete the recipient's immune cells that are reactive against the donor, while preserving other T cell subsets to maintain a robust immune system. This system has the ability to switch targets without reengineering the T cells and respond to multiple antigens to target only alloreactive T-cells and enhance transplantation tolerance. This approach provides a flexible therapeutic platform that can be adapted beyond transplantation to a range of autoimmune diseases including type 1 diabetes, rheumatoid arthritis, vitiligo and other disorders that originate from autoreactive T cell dysfunction.

Virus-Assisted Directed Evolution of Trnas (VADER) in Mammalian Cells

- Conference: Mammalian Synthetic Biology Workshop
- Year: 2021
- Proceeding: 2021 Virtual International Mammalian Synthetic Biology Workshop (mSBW)
- Group: Poster Session
- Session:

  Poster Session

Authors:
Jewel, D. - Presenter, Boston College
Huang, R., Boston College
Kelemen, R., Boston College
Chatterjee, A., Boston College

Site-specific incorporation of non-canonical amino acids (ncAAs) into proteins in living cells, using engineered orthogonal aminoacyl-tRNA synthetase/tRNA pairs, has emerged as a powerful method to probe and engineer protein function. Despite its boundless potential in both in academia and industry, application of this technology in mammalian cells is currently limited by the poor intrinsic performance of the nonsense suppressor tRNA, which interfaces suboptimally with the heterologous translation system of the host cell. Directed evolution, a revolutionary method for the creation of biomolecules with novel and improved activities, has the potential to overcome this limitation. Indeed, improved suppressor tRNAs have been developed for enhanced ncAA incorporation in *E. coli* through directed evolution. However, the ability to perform similar directed evolution experiments in the context of mammalian cells remains poorly developed due to numerous technical challenges. By coupling the activity of the suppressor tRNA to the replication of a human virus, we have developed a platform for 'virus-assisted' directed evolution of tRNAs (VADER) that yielded tRNA variants with remarkably improved suppression efficiency in mammalian cells. This platform is currently being further extended to enable the directed evolution of a broad class of genetic parts in mammalian cells to advance the scope of synthetic biology in these cells.

Control of Hidden Interactions and Competition between Synthetic Gene Circuits and Host Cell

- Conference: Mammalian Synthetic Biology Workshop
- Year: 2021
- Proceeding: 2021 Virtual International Mammalian Synthetic Biology Workshop (mSBW)
- Group: Poster Session
- Session:

  Poster Session

Authors:
Tian, X. - Presenter, Arizona State University

Failure of modularity remains a significant challenge for assembling synthetic gene circuits with tested modules as they often do not function as expected. Hidden circuit-host interactions, such as growth feedback and resource competition, could significantly impair intended circuit function but are often neglected. Our first work revealed a topology-dependent interference of synthetic gene circuit function by growth feedback. Specifically, the memory of the self-activation switch is quickly lost due to the growth-mediated dilution of the circuit products. Decoupling of growth feedback reveals its memory, manifested by its hysteresis property across a broad range of inducer concentration. On the contrary, the toggle switch is more refractory to growth-mediated dilution and can retrieve its memory after the fast-growth phase. Recently, we studied how nutrient level modulates growth feedback and circuit functions and found an unexpected oscillation behavior of a bistable switch circuit. In another work, we unveiled a winner-takes-all resource competition that redirected desired successive cell fate transitions as designed with synthetic cascading bistable switches (Syn-CBS) circuit. We decoupled the resource competition with a division of labor strategy by constructing a two-strain circuit, which achieved successive activation and stable coactivation of the two switches. More recently, we proposed a shared and tunable system of negatively competitive regulation (NCR) and demonstrated NCR significantly increases control efficacy over widely-used global and local negative feedback controllers.

A miRNA-Based NOT Gate As a Building Block of More Sophisticated Circuits in Mammalian Cells

- Conference: Mammalian Synthetic Biology Workshop
- Year: 2021
- Proceeding: 2021 Virtual International Mammalian Synthetic Biology Workshop (mSBW)
- Group: Poster Session
- Session:

  Poster Session

Authors:
Alighieri, G. - Presenter
Weiss, R., Massachusetts Institute of Technology

Scaling up genetic circuits in mammalian cells can lead to a new class of therapies. However, large DNA footprints limit in the choice of safer gene delivery platform like the AAV viral vector, which has a small package capability at about 4.7kb. On the other hand, the use of exogenous proteins can limit the applications in vivo because they can trigger dangerous immune reactions.

Here, we show a miRNA-based NOT gate that has a small DNA footprint and that does not use transcriptional regulators or, more in general, exogenous proteins. This can be the building block of more sophisticated information processing in mammalian cells.

Active miRNAs recognize their target RNA through Watson and Crick base-pairing, which in turn provides a large design space. Additionally input and output signals are biomolecules of the same type. Because of that, miRNA-based NOT gates might be easily combined to make more complex logic circuits.

This NOT gate was implemented with just one transcriptional unit by modifying a ncRNA that naturally shuttles between the nucleus and the cytoplasm. This design successfully showed a NOT-gate behavior in HEK293FT after transiently transfecting the DNA encoding the gate and the input as well.

We are currently investigating the use of the miRNA-based NOT gate for more complex information processing. We are confident that it will push the boundary of what currently can be computed in mammalian cells.

Imaging Cell Lineage with a Synthetic Digital Recording System

- Conference: Mammalian Synthetic Biology Workshop
- Year: 2021
- Proceeding: 2021 Virtual International Mammalian Synthetic Biology Workshop (mSBW)
- Group: Poster Session
- Session:

  Poster Session

Authors:
Chow, K. H. K. - Presenter, California Institute of Technology
Budde, M. W., California Institute of Technology
Granados, A. A., California Institute of Technology
Cabrera, M., California Institute of Technology
Yoon, S., California Institute of Technology
Cho, S., California Institute of Technology
Huang, T. H., California Institute of Technology
Koulena, N., California Institute of Technology
Frieda, K. L., Spatial Genomics
Cai, L., Caltech
Lois, C., Caltech
Elowitz, M. B., California Institute of Technology

During multicellular development, spatial position and lineage history play powerful roles in controlling cell fate decisions. Using a serine integrase-based recording system, we engineered cells to record lineage information in a format that can be read out in situ. The system, termed integrase-editable memory by engineered mutagenesis with optical in situ readout (intMEMOIR), allowed in situ reconstruction of lineage relationships in cultured mouse cells and flies. intMEMOIR uses an array of independent three-state genetic memory elements that can recombine stochastically and irreversibly, allowing up to 59,049 distinct digital states. It reconstructed lineage trees in stem cells and enabled simultaneous analysis of single-cell clonal history, spatial position, and gene expression in Drosophila brain sections. These results establish a foundation for microscopy-readable lineage recording and analysis in diverse systems.

Integrated Circuit Fabrication for Mammalian Cell Biocomputers

- Conference: Mammalian Synthetic Biology Workshop
- Year: 2021
- Proceeding: 2021 Virtual International Mammalian Synthetic Biology Workshop (mSBW)
- Group: Poster Session
- Session:

  Poster Session

Authors:
Letendre, J. - Presenter

Synthetic biological circuits have demonstrated incredible promise for engineering complex, logical control over gene expression. Recently, the Boolean Logic and Arithmetic through DNA Excision (BLADE) single-layer circuit design has enabled the fabrication of highly complex circuits in mammalian cells using site-specific recombinases (SSRs). While SSR-based logic circuits are in valuable for cell-based computation, their digital (on or off) behavior lacks the ability to finely tune gene expression levels. Alternatively, programmable transcription factors offer user-defined gene expression levels but their implementation into circuits has thus far been limited to 2-input logic gates. To bridge the gap between programmability and logic processing, we present the Computation via Recombinase Assisted Transcriptional Effectors (CREATE) platform for multi-input-multi-output (MIMO) integrated circuit design. CREATE uses transcription factors as molecular wiring between layers of BLADE circuits, and has been optimized for use with both dCas9 and zinc finger transcriptional activators. We have established design rules and an extensive promoter library for the facile construction of multiple CREATE circuits in mammalian cells. To demonstrate the power of this design, we present two of the most complex mammalian gene circuits to date: several 2-input digital-to-analog converters (DACs) and the first 4-input 12-output genetic multiplier circuit. CREATE provides anew suite of tools for improved MIMO biocomputation, and will be useful for multiple research applications including the interrogation of complex diseases and drug screening involving multi-pathway regulation.

Reducing Loss of AAV Inverted Terminal Repeats from Transfer Vector Plasmids Using the E. coli SSB Protein

- Conference: Mammalian Synthetic Biology Workshop
- Year: 2021
- Proceeding: 2021 Virtual International Mammalian Synthetic Biology Workshop (mSBW)
- Group: Poster Session
- Session:

  Poster Session

Authors:
Chanas, S. - Presenter, GlaxoSmithKline

Transient productions of recombinant AAV vectors in mammalian cells utilise a plasmid carrying a transgene flanked by AAV inverted terminal repeats (ITRs). ITRs are required for replication of the transfer vector, subsequent encapsidation and stability of the recombinant genome in transduced cells. During replication in *E. coli,* a proportion of these plasmids (~5-100%) suffer a partial deletion of an ITR, resulting in heterogeneous preparations; a universal problem in AAV vectorology. Recombinant AAV genomes flanked by ITRs containing deletions are inefficiently packaged into the capsid in mammalian cells, reducing target vector yields and giving unpredictable *in vivo* performance. These ITR deletions were postulated to be driven by formation of hairpin structures in the lagging strand of the DNA replication fork. In this model, the hairpin is bridged by an Okazaki fragment, resulting in daughter plasmids lacking part of the ITR sequence. We hypothesised that overexpression of the *E. coli* single stranded binding protein (SSB), that binds all single stranded DNA in the bacterium, would prevent hairpin formation. Cloning the *ssb* gene into an AAV transfer vector plasmid backbone so that SSB protein would be overexpressed in transformed bacteria resulted in greatly reduced ITR loss, potentially improving vector quality. Other factors may contribute, but addition of *ssb* to AAV transfer vector plasmids has largely solved the ITR instability issue, eased cloning steps for new AAV gene therapies and will potentially improve quality and yield of vector.

Resource-Aware Construct Design in Mammalian Cells

- Conference: Mammalian Synthetic Biology Workshop
- Year: 2021
- Proceeding: 2021 Virtual International Mammalian Synthetic Biology Workshop (mSBW)
- Group: Poster Session
- Session:

  Poster Session

Authors:
Di Blasi, R. - Presenter, Imperial College London
Tedeschi, F., Istituto Italiano Di Tecnologia (IIT)
Pisani, M., Istituto Italiano di Tecnologia (IIT)
Furini, S., University of Siena
Ceroni, F., Imperial College London
Siciliano, V., Italian Institute of Technology

Despite recent advances in mammalian synthetic biology, the design of predictable and robust synthetic constructs still proves challenging. The competition for intracellular resources has been reported as one relevant cause of circuit failure in both bacterial and mammalian cells, and mitigation strategies to address this problem are starting to emerge (Ceroni *et al.*, 2015; Frei *et al.*, 2020; Jones *et al.*, 2020; Di Blasi *et al.*, 2021).

While in bacteria the comparison among different construct design parameters (e.g. promoter strength, RBS) has led to more robust expression over time, in mammalian cells such an analysis has not been performed yet, precluding potential improvement of mammalian synbio applications.

Here, we investigate the rules for efficient circuits assembly by testing the composition of different design parameters in a panel of mammalian cell lines. In particular, we show how tuning promoter strength, Kozak and polyadenylation efficiency results in designs that place a different demand on cellular resources when transiently expressed. We also test the secretion system as further bottleneck in the context of resource shortage.

This study lays initial foundations to identify new design rules predictive of reduced resource competition in mammalian cells, leading the way towards a more robust cell engineering that can benefit both biotherapy and bioproduction.

Proportional-Integral Feedback Controllers for Robust and Tight Mitigation of Mammalian Gene Expression Burden

- Conference: Mammalian Synthetic Biology Workshop
- Year: 2021
- Proceeding: 2021 Virtual International Mammalian Synthetic Biology Workshop (mSBW)
- Group: Poster Session
- Session:

  Poster Session

Authors:
Frei, T. - Presenter, ETH Zürich
Chang, C. H., ETH Zürich
Filo, M., ETH Zürich
Khammash, M., ETH Zürich

The modular engineering of complex genetic circuits is often complicated by unexpected interactions between individual circuit components. Recently, the competition for finite pools of shared resources in gene expression has been identified as a major source of such interactions. Here, we introduce a mammalian implementation of antithetic integral and proportional-integral feedback to robustly control gene expression in various contexts. To realize integral feedback, we exploit the sequestration of the mRNA of a synthetic transcription factor by its antisense complement and use it to compute the mathematical integral of the error between a reference and the transcription factor expression level. The resulting integral feedback control robustly and precisely matches the expression level of the transcription factor to the reference. By introducing additional negative feedback from the mRNA of the synthetic transcription factor on its translation through a bicistornically encoded RNA-binding protein, the integral feedback is extended to realize proportional-integral feedback. We show that the integral controller rejects disturbances that result from increasing the degradation rate of the controlled transcription factor while providing adaptation to changes in the structure of the controlled network. We then show that the proportional-integral controller, in addition, reduces cell-to-cell variability of the controlled species. Finally, we introduce supplementary expression of a constitutive fluorescent protein as a disturbance to the availability of gene expression resources and show that both the integral and proportional-integral controllers manage to maintain the expression of the synthetic transcription factor. Again, the proportional-integral controller also demonstrably reduces the variability of the expression of the synthetic transcription factor in comparison to the integral controller alone. We expect that integral and proportional-integral controllers are going to be important additions to the synthetic biologist's toolkit, adding robustness and precision to cellular function through more sophisticated negative feedback control.

Rational Design of a Bifunctional and-Gate Ligand to Modulate Cell-Cell Interactions

- Conference: Mammalian Synthetic Biology Workshop
- Year: 2021
- Proceeding: 2021 Virtual International Mammalian Synthetic Biology Workshop (mSBW)
- Group: Poster Session
- Session:

  Poster Session

Authors:
Lee, J. - Presenter
Way, J. C., Harvard University

Much of synthetic biology focuses on the engineering of gene expression systems, using linear relationships between nucleic acid sequences and output proteins, plus simple pathway relationships that can be modeled by one-dimensional interactions. In contrast, Nature is three-dimensional, and evolution uses spatial geometry to construct optimized systems. The molecular geometry of domains in engineered fusion proteins is crucial for designing artificial synthetic-biological systems, such as chimeric antigen receptor T cells and bispecific T cell engagers. Of particular interest are fusion proteins composed of two or more domains that deliver a therapeutic activity to target cells while minimizing undesired action on other cells that may lead to side effects. Implicit in this design is the potential to crosslink cells, whether desired or not. Here, we constructed a bifunctional AND-gate protein, which activates a subset of cells that bear two different receptors and avoids action on other cells. These experiments address how protein structure and geometry can inform the rational design of a synthetic system. Specifically, erythropoietin (EPO) was directed to red blood cell (RBC) precursors by fusion to an anti-glycophorin A (GPA) antibody V region to stimulate RBC production without causing thrombosis. Several different forms of anti-GPA-EPO fusion proteins stimulated RBC and not platelet production in mice. However, they also caused adhesion between RBCs and EPO receptor-bearing cells in vitro and enhanced thrombosis in mice. Based on the protein structural model of the RBC surface, we rationally designed a form of an anti-GPA-EPO fusion protein that does not mediate cell-cell adhesion or enhance thrombosis in mice. We found that the linker element of the fusion protein, the position of the antibody binding epitope, and the target cell surface proteins synergistically contribute to cell-cell interaction mediated by the fusion protein. This work shows how synthetic biology can use nano-scale three-dimensional engineering.

Optogenetic Control of Store-Operated Calcium Channels

- Conference: Mammalian Synthetic Biology Workshop
- Year: 2021
- Proceeding: 2021 Virtual International Mammalian Synthetic Biology Workshop (mSBW)
- Group: Poster Session
- Session:

  Poster Session

Authors:
MA, G. - Presenter, Texas A&M University
He, L., Texas A&M University
Zhou, Y., Texas A&M University

Store-operated Ca^{2+} entry mediated by STIM1 and ORAI1 constitutes one of the major Ca^{2+} entry routes in mammalian cells. Physiologically, the molecular choreography of STIM1–ORAI1 coupling is initiated by endoplasmic reticulum (ER) Ca^{2+} store depletion with subsequent oligomerization of the STIM1 ER-luminal domain, followed by its redistribution towards the plasma membrane to gate ORAI1 channels. Over the past five years, we have engineered a series of genetically-encoded Ca^{2+} actuators (GECAs) derived from STIM and ORAI to enable optical control of Ca^{2+} signaling with unprecedented spatial and temporal resolution. STIM1 is an ER Ca^{2+} sensor that oligomerizes and conformational switch in response to store depletion, overcoming autoinhibition and forming clusters that then couples to ORAI channel. Two general engineering approaches have been applied to confer photosensitivity to STIM1: i) fusion of a photosensory module with SOAR/CAD to recapitulate reversible CC1-SOAR/CAD intramolecular trapping (Opto-CRAC-V1); ii) replacing the STIM1 luminal domain and the transmembrane domain with a light-inducible oligomerization domain (Opto-CRAC-V2). These tools enable light-dependent STIM-ORAI coupling to elicit Ca^{2+} influx through highly Ca^{2+}-selective CRAC channels in both excitable and non-excitable cells.

Meanwhile, we designed a light-operated Ca^{2+} channel (designated LOCa) by inserting a plant-derived photosensory module into the intracellular loop of an engineered ORAI1 channel. LOCa displays biophysical features reminiscent of the ORAI1 channel, which enables precise optical control over Ca^{2+} signals and hallmark Ca^{2+}-dependent physiological responses. Compared to traditional pharmacological or genetic tools, GECAs rival by high precision, rapid reversibility, facile tunability and non-invasiveness. These GECAs offer new non-invasive means to probe the structure-function relations of calcium channels and significant advances toward the use of optogenetics for immunomodulation and neuromodulation.

Development of a Synthetic Genetic Feedback Circuit to Sense and Modulate Neurodegenerative Protein Aggregation

- Conference: Mammalian Synthetic Biology Workshop
- Year: 2021
- Proceeding: 2021 Virtual International Mammalian Synthetic Biology Workshop (mSBW)
- Group: Poster Session
- Session:

  Poster Session

Authors:
Chiesa, G. - Presenter, Boston University
Webber, C., Boston University
Khalil, A. S., Harvard University
Wolozin, B., Boston University School of Medicine

Protein aggregation is a pathological hallmark of many neurodegenerative disorders, such as Alzheimer's Disease, that affect an expanding proportion of the human population[1]. Despite the significant burden caused by protein aggregation diseases, there are currently no effective therapies that address the underlying pathology. Recent efforts have focused on modulating common molecular mechanisms, such as enhancing proteostasis, or inhibiting protein aggregation[2]. Here, we propose an alternative therapeutic solution: an adaptive gene therapy based on a synthetic feedback circuit that senses pathological aggregation of a target protein, and responds by expressing an aggregation-modulatory "therapeutic" gene.

The design of our genetic circuit is based on aggregation-dependent reconstitution of a split TEV protease, a well-established genetic system, recently used to engineer complex protein-based circuits in mammalian cells[3]. Here, we fuse each subunit of a split TEV protease to the aggregation-prone region of a disease-relevant protein. Protein assembly/aggregation leads to TEV reconstitution, the specific cleavage of a degron and subsequent activation of a therapeutic protein.

To validate circuit components, we first demonstrate sensing of TDP43 and Tau aggregation, two proteins critically involved in Alzheimer's disease, in HEK293T and SH-SY5Y neuronal cell lines. Second, using a cellular model of Tau pathology, we have developed a screen to identify potential therapeutic genes, capable of modulating pathological aggregation in cells. Our cell-autonomous genetic circuit can be encoded on a single viral vector, and may facilitate the development of next-generation gene therapies for neurodegeneration.

1. Hou, Y. *et al.* Ageing as a risk factor for neurodegenerative disease. *Nature Reviews Neurology* (2019). doi:10.1038/s41582-019-0244-7
2. Scannevin, R. H. Therapeutic strategies for targeting neurodegenerative protein misfolding disorders. *Curr. Opin. Chem. Biol.* **44**, 66–74 (2018).
3. Gao, X. J., Chong, L. S., Kim, M. S. & Elowitz, M. B. Programmable protein circuits in living cells. *Science (80-.).* **361**, 1252–1258 (2018).

AUTHOR INDEX